KV-280-053

CONTENTS

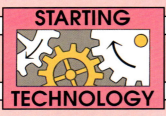

FLYING MACHINES

For thousands of years people have wanted to fly. They looked at birds and wondered how they were able to move through the air.

Today, huge **jet aeroplanes** with hundreds of **passengers** on board travel all around the world. Have you ever flown in an aeroplane?

A modern jet aeroplane like this one can travel thousands of kilometres without having to stop for fuel.

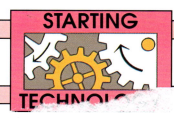

FLIGHT

John Williams

Illustrated by
Malcolm S. Walker

Titles in this series

AIR
COLOUR AND LIGHT
ELECTRICITY
FLIGHT
MACHINES
TIME
WATER
WHEELS

Words printed in **bold** appear in the glossary on page 30.

© Copyright 1991 Wayland (Publishers) Ltd

First published in 1991 by
Wayland (Publishers) Ltd
61 Western Road, Hove
East Sussex BN3 1JD, England

Editor: Anna Girling
Designer: Kudos Design Services

British Library Cataloguing in Publication Data
Williams, John
  Flight.
  1. Flight
  I. Title II. Series
  629.132

HARDBACK ISBN 0-7502-0026-X

PAPERBACK ISBN 0-7502-0918-6

Typeset by Kudos Editorial and Design Services, Sussex, England
Printed in Italy by Rotolito Lombarda S.p.A.
Bound in Belgium by Casterman S.A.

Finding out about flying

The aeroplane is a fairly modern **invention**. The first aeroplane with an engine to power it through the air was flown just over eighty years ago. It did not look at all like a modern aeroplane. It was made of canes, cloth and wire. Its flight lasted only a few seconds.

Before this, people had flown in **hot-air balloons** and **gliders**, but never in a powered aeroplane.

1. Try to find out about the history of flying. Draw pictures of hot-air balloons, gliders and the first aeroplanes. Cut out pictures of different modern aircraft. Stick your drawings and pictures in a scrapbook.

2. Watch the birds near your home. Look at their feathers and the shape of their wings. Draw pictures to show how they move their wings as they fly.

3. Perhaps you have flown in an aeroplane. Write a story about what happened before and during your flight.

PAPER DARTS

Making a simple paper dart

You will need:

A piece of A4 paper

Fold the paper along the dotted lines, carefully following these drawings.

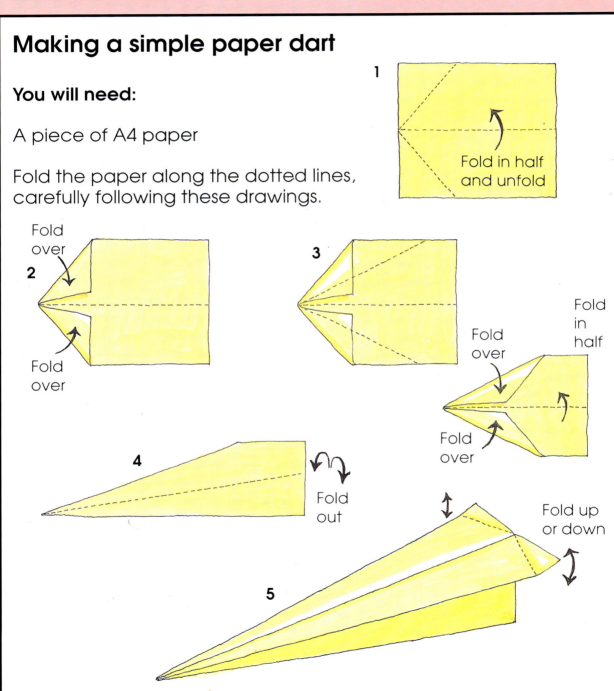

1 Fold in half and unfold

2 Fold over / Fold over

3 Fold over / Fold over / Fold in half

4 Fold out

5 Fold up or down

Testing your paper dart

Throw your dart from shoulder height. Never throw it too hard. A gentle throw works better.

The flaps at the ends of the wings will help to steer the dart. Fold one flap up at a time and see which way the dart turns. Fold one flap down at a time and see if this makes any difference. What happens to the dart when you throw it with one flap up and one flap down?

Concorde can fly faster than the speed of sound. Look at its shape. Are your paper darts similar?

GLIDERS

The paper dart you made on page 6 was a very simple glider. Gliders do not have engines to push them through the air. Once they have been **launched** they use their wings to stay in the air and come gently down to the ground.

Gliders are very strong and light. The long, thin wings are specially designed to help them fly for as long as possible.

Making a glider

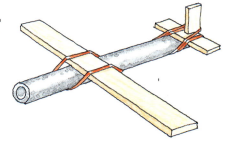

You will need:

Two pieces of balsawood, about 8 cm wide
 and 60 cm long
A junior hacksaw
A piece of plastic foam tubing, about 4 cm in
 diameter and 70 cm long (from a DIY shop)
Rubber bands
A ruler

1. For the wings, fix a piece of balsawood
across the foam tubing with a rubber band.

2. From the other piece of balsawood, cut
two pieces, one 30 cm long and the other
20 cm long.

3. Measure the edge of the longest piece
and find its central point. At this point cut a
slit 2 cm into the wood.

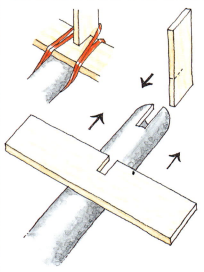

4. Fix this piece of wood to the tail end
of the foam tube with a rubber band.
The slit should face towards the back.

5. Make another slit 3 cm long in the
top of the foam tube, at the tail end.

6. Use the shorter piece of wood as the
rudder. Slide it into the slit in the tube
and fix it into the slit in the other piece of
wood. Hold it in place with a rubber
band.

GOING GLIDING

Flying your glider

Before you fly your glider you need to make sure it is properly balanced. Prop up your glider by resting the tips of the wings on the backs of two chairs. If the tail end drops, the glider is tail-heavy. To correct this, put some plasticine on the nose of the glider. If it is nose-heavy put some plasticine on the tail.

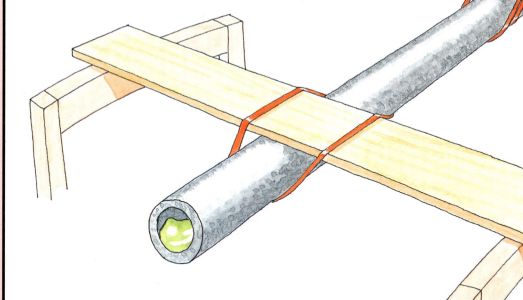

If necessary, slide the wings down the body of the glider. They should not be too near the nose.

Now find an open space to throw your glider and see how it flies.

Further work

Try your glider outside. Does the strength of the wind make a difference to how it flies? Does your glider fly better with the wind behind it or against it? **Measure** how far it flies.

Look at this hang-glider. Could you make a model of it?

HELICOPTERS

When aeroplanes take off they have to build up speed on a **runway** before they can start flying. Helicopters do not need runways because they can take off straight upwards. They have **rotor blades** on them. When the blades spin round they lift the helicopter off the ground.

A helicopter has rotor blades instead of wings to keep it in the air. Can you see the blades spinning round?

Making a paper helicopter

You will need:

Paper
Scissors
A paperclip

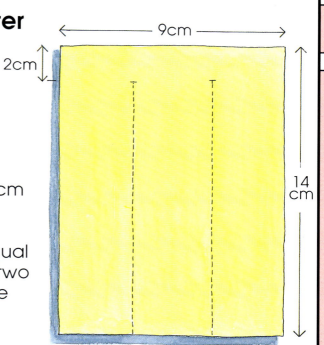

1. Cut out a piece of paper 9 cm wide and 14 cm long.

2. Fold the paper into three equal strips. Unfold the paper leaving two creases running the length of the paper.

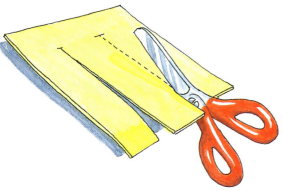

3. Cut along the two creases. Stop cutting about 2 cm from one end.

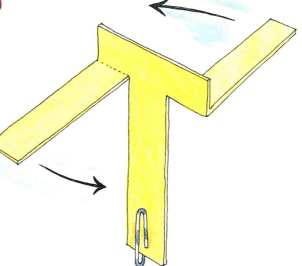

4. Fold one of the outside strips backwards and the other forwards. These are the rotor blades of your helicopter. Attach a paperclip to the bottom of the middle strip of paper.

5. Hold your helicopter above your head and let it drop through the air. Watch it spin to the ground.

Making a balsawood helicopter

You will need:

Two sticks of balsawood,
 about 12 cm long
Card
Scissors

Glue
A plastic bead
A pin

1.   Cut out two pieces of card, 3 cm by 2 cm.

2.   Glue the pieces of card to a stick of balsawood, one at each end. The pieces of card should overlap the wood on opposite sides. Fold up the overlaps to make flaps. This is now your rotor blade.

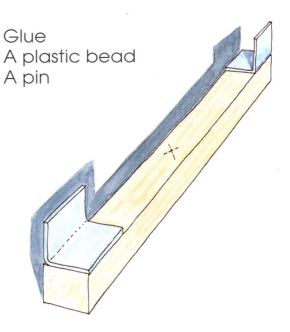

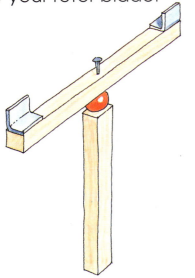

3.   Push the pin through the centre of the rotor blade, then the bead and finally into one end of the other balsawood stick. Make sure the rotor blade can spin freely on the bead.

4.   Hold your helicopter in the air and let it drop — just like you did with the paper helicopter.

Further work

Try bending the pieces of card into different angles on your rotor blade. Does it make a difference to the way the blade spins?

You can make a pilot for your helicopter. Make a seat out of card and fix it to the bottom of your helicopter with rubber bands. Make a small model person out of plasticine and pipe cleaners and attach your pilot to the seat with more rubber bands. Will your helicopter still work?

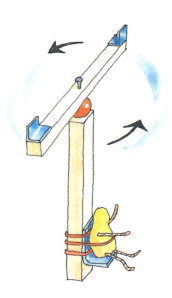

Helicopters do not need runways for taking off and landing. This helicopter is landing on an oil rig far out to sea.

NATURE'S SPINNERS

Many plants produce seeds from which new plants grow. When these seeds fall from the plant they may travel a long way before they reach the ground. This stops all the new plants growing in the same place. Some seeds have clever ways of travelling further. Some have wings which make them spin round — rather like a helicopter's rotor blades. The seeds of sycamore trees are like this.

Look at the wings on these sycamore seeds. Try to find other trees and plants that have seeds which can fly or float through the air.

Experimenting with seeds

You will need:

Sycamore seeds
String
Chalk
A chair

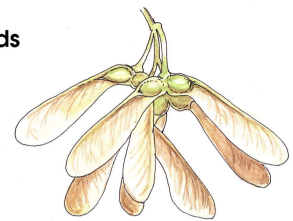

1. Collect as many sycamore seeds as you can.

2. Draw several circles in your school playground, one inside another. To do this, tie some chalk to a piece of string. Get a friend to hold the end of the string at one point on the ground while you draw round with the chalk. The diameter of each circle should be a metre larger than the one before.

3. Stand on a chair in the middle of the circles. **Make sure there is an adult with you.** Drop your seeds one at a time. How far do they fly before they reach the ground? Does it make a difference if there is a wind blowing?

4. Test your paper helicopter in this way. Does it work as well as the sycamore seeds?

MINI-GLIDERS

Leonardo da Vinci was a famous scientist and artist who lived more than 500 years ago. He tried to invent a flying machine — long before anyone had dreamed it was possible. Leonardo **designed** and built several model gliders. He had no other designs to copy, so he looked at birds for ideas.

This seagull is gliding through the air. The shape of its wings helps it to fly.

Making a Leonardo da Vinci-style mini-glider

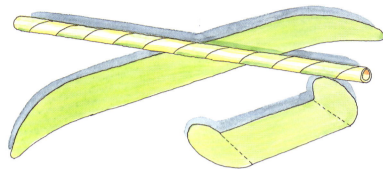

You will need:

Plastic straws
Stiff paper
Scissors
Sticky tape
Plasticine

1. Cut out wings and a tail from paper. Try to make them look like the shape of a bird's wings and tail.

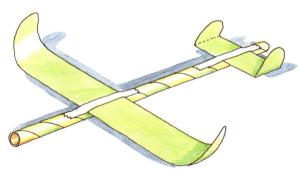

2. Bend up the ends of the wings and tail. Stick them on to a straw with sticky tape.

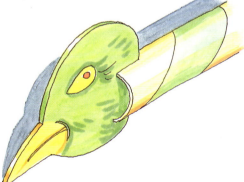

3. Draw a bird's head on some paper, cut it out and attach it to the front of your mini-glider.

4. Fix a little piece of plasticine to the front of your glider so that it will balance. Now try flying it.

5. You can make other mini-gliders in this way. Look at different birds and copy them to make new designs.

FAIR TESTS

When you and your friends have made lots of paper darts, gliders and mini-gliders, you will need to test them. You could try throwing them, but one of you may be stronger than the rest and so be able to throw his or her dart further. This would not be fair!

This child is about to throw a paper dart. Will it go as far as the ones you made?

Making a catapult for a paper dart

You will need:

A short plank of wood   A pencil
Two small nails   Scissors
A small hammer   Sticky tape
A rubber band

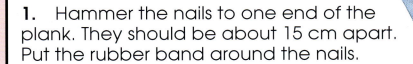

1. Hammer the nails to one end of the plank. They should be about 15 cm apart. Put the rubber band around the nails.

2. Draw a line across the plank, about 40 cm from the rubber band.

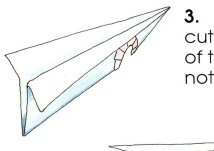

3. To catapult a paper dart, you will need to cut a notch in the under-side, near the nose of the dart. Strengthen the paper around the notch with sticky tape.

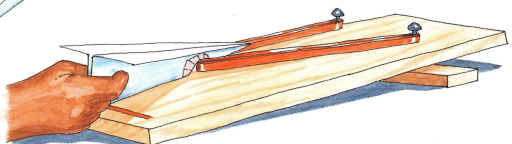

4. Put the rubber band into the notch and pull the dart back until its tail is level with the line. Let the dart go and watch it fly.

5. If all the darts are flown in this way the test will be fair. Whose dart flew the furthest? Try testing your gliders and mini-gliders with your catapult too.

WARNING: Ask an adult to help you use the hammer and nails.

BIRDS

Making a mechanical bird

WARNING: Ask an adult to help you cut balsawood.

You will need:

A sheet of thin balsawood
A junior hacksaw
Glue
Pencils
String
Sticky tape
A wooden stick
Plasticine

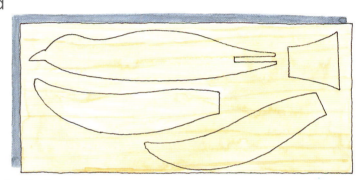

1. Draw the shape of a bird's body, wings and tail on the balsawood and cut them out with the hacksaw. The bird should be about 40 cm long, including the tail. The wings should each be about 20 cm long.

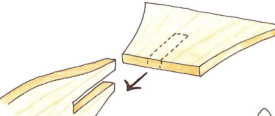

2. Cut a slit in the end of the bird's body. Stick the tail in the slit with glue.

3. Fix the wings to the centre of the body with sticky tape. They should be about 15 cm from the beak.

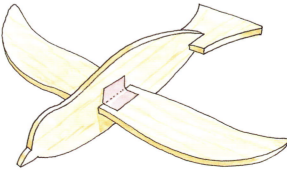

4. With more tape, attach a piece of string to the middle of each wing. Tie the strings to the stick. Attach another piece of string to the underside of the bird.

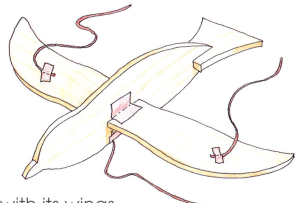

5. To make the bird balance with its wings out, stick small pieces of plasticine to the head and wing tips and a larger piece to the string underneath. You will need to experiment with different amounts of plasticine to find the correct amount.

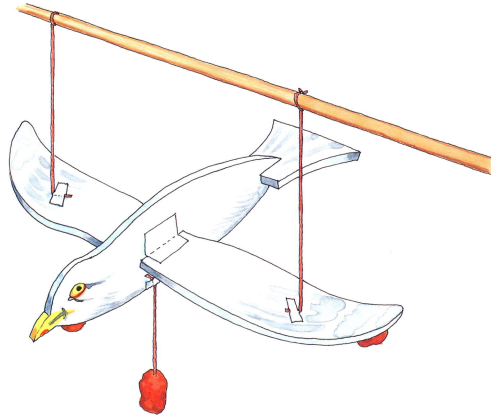

6. When the bird is balanced, pull the string underneath gently and the wings will flap.

KITES

Have you ever flown a kite in the park or at the seaside? Kites can be many different shapes and sizes, but they are all shaped so that the air will lift them up and make them fly.

Making a kite

You will need:

A plastic refuse sack
Two garden canes,
  about 90 cm long
Sticky tape
Scissors
String

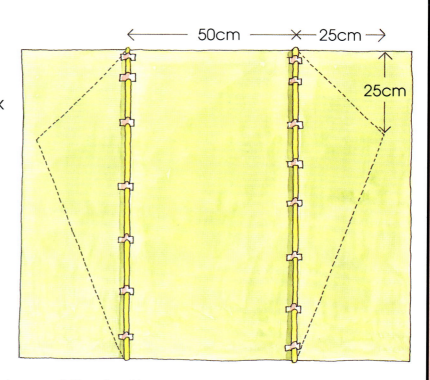

1. Cut open one side and the bottom end of the plastic sack. Open it up and lay it out on the floor.

2. Lay the two garden canes on the sack, 50 cm apart. Leave the same amount of sack on each side of the canes.

3. Fix the canes on to the sack with sticky tape. Make sure the ends of the canes are firmly fixed.

4. Cut two triangular side flaps from the sack, following the diagram on the opposite page to get the right shape.

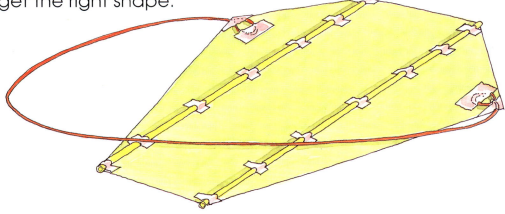

5. Cut a piece of string 4 m long and tie a little loop at each end. This will be the **bridle**. With sticky tape, stick the loops to the triangular side flaps. Make sure the loops point towards the centre of the kite.

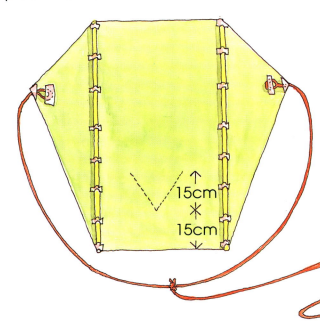

15cm
15cm

6. Make a small V-shaped cut in the kite, as in this diagram. When the kite is flying, the cut will open to let air through. This helps the kite to stay level.

7. Tie a long piece of string to the bridle string.

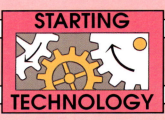

KITE-FLYING

Flying a kite

You will need a friend to help you fly the kite shown on pages 24 and 25.

1.   Let out about 50 m of string.

2.   Get your friend to hold the kite high in the air. As soon as the kite starts to fly your friend must let it go.

3.   If the wind is quite strong you will be able to stand still holding the string, letting out more string as the kite flies higher. If there is not much wind you may need to run with your kite to make it fly.

WARNING: Never fly a kite where there are electric cables, or near a road or railway line.

Further work

You can measure how strong the pull of your kite is. When it is flying in the air, attach the end of the string to the hook of a spring balance. Make sure you keep hold of the ring on the other end of the spring balance. Look to see where the marker is on the scale. Measure the pull on different days. Does it vary?

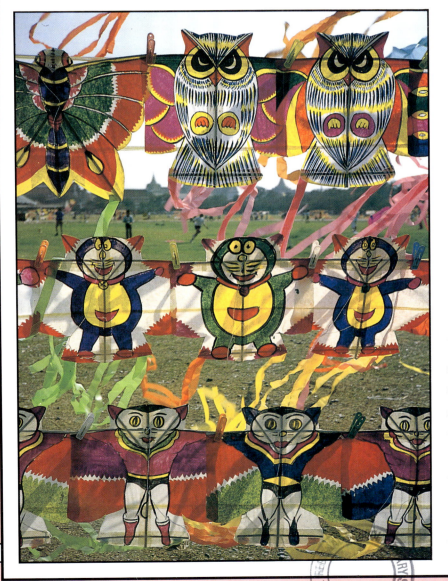

These kites have colourful pictures on them. What kind of decoration would you like to have on your kite? You could try painting some pictures of your own.

SCIENCE AND TECHNOLOGY

Flight is a popular topic in primary schools. Children will delight in making the various darts, gliders and other models. However, they should also be given an opportunity to develop ideas about how an aeroplane or glider flies and how a kite stays in the air.

There are four forces involved in flight. Lift is the force that holds a craft in the air against gravity. Gravity is the force pulling the craft down. Thrust is the force that pushes or pulls it along, and drag is the resistance exerted by the air on an object moving through it. If the lift produced by a wing is greater than gravity, and if the thrust produced by forward motion is greater than drag, then the craft will fly.

This is, of course, a very simple explanation. For children of this age complex details and ideas are not necessary. As they make the models children will begin to understand what is necessary for flight, and how to develop their models to make them fly better.

All the models in this book have been made by young children. Some help may be needed at times, particularly by the younger children and less able readers. However, children should be encouraged to carry out as much of the work as possible for themselves. They should develop their own ideas and come up with variations on the basic designs in this book.

Safety should be stressed at all times. Junior hacksaws, scissors, pliers and small hammers are the tools recommended and, when properly used, are quite safe. Balsawood knives are **not** recommended.

HISTORY

Any study of flight will involve its history. Children can learn about the first hot-air balloons, early attempts at powered flight and the airships of the 1920s and 1930s. They can also be introduced to the mythology of flight — for example, the story of Icarus, or even magic carpets. The topic of flight might also be part of a larger project about transport.

MATHEMATICS

All testing of models involves measurement and therefore uses basic mathematics. The recording of these tests can involve making scale diagrams, graphs and charts.

LANGUAGE

Language skills should develop from children discussing their activities. Children can also write about their work and should be encouraged to do so in an imaginative way.

National Curriculum Attainment Targets

This book is relevant to the following Attainment Targets in the National Curriculum for science:

Attainment Target 1 (Exploration of science)   The construction, development and testing of the models throughout this book is relevant. Particular attention should be paid to the development of a fair test (level 3).

Attainment Target 10 (Forces)   This Attainment Target is the one most involved in a study of flight. Forces are at work in pushing or pulling flying objects through the air, while friction (drag) tries to hold them back and gravity tries to keep them on the ground.

Attainment Target 13 (Energy)   All the working and moving models in this book involve some form of energy.

The following Attainment Target is included to a lesser extent:

Attainment Target 2   (Variety of life) The topic of flight will involve the study of birds and flying insects. Work on seeds in this book involves the study of plant life.

Teachers should also be aware of the Attainment Targets covered in other National Curriculum documents — that is, those for design and technology, mathematics, history and language.

Bridle   A Y-shaped piece of rope used to hold on to something.

Design   To draw a plan for making something.

Diameter   The length of a straight line going through the centre of a circle, from one side to the other.

Gliders   Aircraft that do not have engines.

Hot-air balloons   Large balloons filled with hot air which lifts them off the ground.

Invention   The creation of a new machine.

Jet aeroplanes   Aeroplanes powered by jet engines. A jet engine produces a jet of gas to thrust the aeroplane forwards.

Launch   To send an aircraft into the air or a boat into the water.

Measure   To find out how big, heavy or long something is.

Passengers   The people travelling in an aeroplane, car, bus or train, except the pilot or driver.

Rotor blades   The flat, metal 'arms' that spin round on top of a helicopter and lift it off the ground.

Rudder   A flat piece fixed to the back of an aeroplane and used for steering it.

Runway   A long flat piece of land used by aeroplanes for taking off and landing.

BOOKS TO READ

Air in Action   by Robin Kerrod (Cherrytree, 1988)
Catching a Plane   by Brenda Clark (Kingfisher, 1988)
The First Flyers   by David Jefferis (Franklin Watts, 1988)
The First Transatlantic Flight   by Mike Rosen (Wayland, 1989)
Helicopters   by Ian Graham (Franklin Watts, 1989)
In the Air   by Mary Gribbon (Macdonald, 1987)
Let's Look At Aircraft   by Andrew Langley (Wayland, 1989)

Picture acknowledgements
The publishers would like to thank the following for allowing their photographs to be reproduced in this book: Heather Angel 16; Cephas Picture Library 12; Eye Ubiquitous 4, 27; Hutchison Library 15, 18; Topham Picture Library 8; Timothy Woodcock 20; Zefa 7, 11. Cover photography by Zul Mukhida.

INDEX